J. WESTON
WALCH
PUBLISHER
Portland, Maine

Content-Area Reading Strategies

Science

Gina Hamilton

User's Guide
to
Walch Reproducible Books

Purchasers of this book are granted the right to reproduce all pages.

This permission is limited to a single teacher, for classroom use only.

Any questions regarding this policy or requests to purchase further reproduction rights should be addressed to

>Permissions Editor
>J. Weston Walch, Publisher
>321 Valley Street • P.O. Box 658
>Portland, Maine 04104-0658

1 2 3 4 5 6 7 8 9 10

ISBN 0-8251-4574-0

Copyright © 2003
J. Weston Walch, Publisher
P.O. Box 658 • Portland, Maine 04104-0658
walch.com

Printed in the United States of America

Contents

Introduction: To the Teacher v

Part 1: Building Vocabulary 1
    Lesson 1: Using Context Clues 2
    Lesson 2: Word Parts 5
    Lesson 3: Word Groups 7

Part 2: Prereading 9
    Lesson 4: Previewing 10
    Lesson 5: Predicting 12
    Lesson 6: Prior Knowledge 13
    Lesson 7: Purpose 14

Part 3: Reading Strategies 15
    Lesson 8: Introduction to Reading Strategies 16
    Lesson 9: KWL 18
    Lesson 10: SQ3R 22
    Lesson 11: Semantic Web 27
    Lesson 12: Outline 32
    Lesson 13: Structured Notes 37

Part 4: Postreading 41
    Lesson 14: Summary and Paraphrase 42

Part 5: Reading in Science 47
    Lesson 15: Common Features and Patterns in Science Reading 48
    Lesson 16: Special Terms 49
    Lesson 17: Topic and Subtopic 51
    Lesson 18: Classification 53
    Lesson 19: Steps in a Process 55
    Lesson 20: Assertion and Support 57
    Lesson 21: Review 59

Part 6: Practice Readings 61
    Reading A: A Year Around the Sun 62
    Reading B: The Strange Creatures of the Burgess Shale ... 65
    Reading C: What Is a Cell? 67

Blank Graphic Organizers 71
    4-P Chart 72
    KWL Chart 73
    SQ3R Chart 74
    Semantic Web 75
    Outline 76
    Structured Notes 77

Teacher's Guide and Answer Key 79

INTRODUCTION
To the Teacher

The *Content-Area Reading Strategies* series teaches students how to read to learn. In the early grades, students learn to read and write narratives—stories. They are used to dealing with texts that have a beginning, a middle, and an end. They expect to encounter rising action that leads to a climax and then to a resolution.

This pattern of organization is often not followed in informational texts, which begin to make up a large portion of classroom-related reading about grade four. Without instruction in how to read these kinds of nonnarrative texts, even good readers can stumble. Some research shows that the so-called fourth-grade "slump" may be attributable in part to unsupported transition from narrative to informational texts.

That's where the *Content-Area Reading Strategies* series comes in. Each book in the series focuses on a different content area and gives students concrete tools to read informational texts efficiently, to comprehend what they read, and to retain the information they have learned.

Organization is an important part of comprehending and retaining knowledge. The graphic organizers in *Content-Area Reading Strategies* help students connect new information to their existing schemata, thus increasing their ability to recall and take ownership of what they read. The reading strategies give students a way to "see" what they read—a great asset to visual learners.

The reading-writing connection is a strong one. The reading strategies in this book all require students to record information in writing, strengthening readers' ability to retain and access newly acquired knowledge.

The "think along" (modeling) strategy used throughout this book should help your students begin using their own considerable knowledge as a tool in tackling new and challenging subject matter. Often, especially in middle school, students need to discover how much background information they already possess about a subject. Thinking aloud is an extremely helpful approach to decoding and assimilating new data.

The science topics chosen for this book address concepts in the national science standards for middle school students. The book is not, however, designed as a science curriculum. Instead, our aim is to help students learn to develop reading strategies within this content area, and to critically analyze information. These tools will help them not only succeed in science class, but become better informed in daily life issues as well.

Classroom Management

Content-Area Reading Strategies is easy to use. Simply photocopy each lesson and distribute it. Each lesson focuses on a single strategy and includes models showing the strategy in action. Blank copies of each graphic organizer are included at the back of the book so you can copy them as often as needed. Quiz questions assess how well students understand what they read.

The practice readings provide longer readings and questions. For these, you may want to let students choose which strategy to use, or you may assign a particular strategy. Either way, have copies of the appropriate graphic organizers available.

Eventually, students will no longer need printed graphic organizers; they will make their own to suit their own learning style and the particular text they are reading. They will have integrated the reading strategies as part of the learning process in all content areas.

PART 1
Building Vocabulary

Lesson 1
Using Context Clues

Vocabulary can be one of the hardest things about any new subject area. Especially in science, you may see words you never have seen before. Since any subject has vocabulary that is specific to that subject, a general understanding of these words is important. This lesson, and the next two lessons, will give you some ways to build vocabulary while you read.

Analyzing Context Clues

Pick up clues from the context of what you are reading. Authors try to give vocabulary help to readers, especially when they know the readers might not be familiar with all the terms. How do they do this?

- **Definitions.** An author may define a word, or provide a *synonym*—a word that has the same meaning as the first word. Let's say you read the following sentence: "Rain forest (jungle) plants have roots near the surface of the soil." The author tells you that another word for "rain forest" is "jungle."

- **Opposites.** Sometimes a new word is the opposite (*antonym*) of a common word. Look at this sentence: "Unlike the adult insect, the larval mosquito lives in water." Here, the author tells us about a difference between the adult and the larval forms. We can guess, from context, that the larval mosquito is a juvenile—the opposite of an adult.

- **Examples.** Often, the best way to define a thing is to provide examples. For instance, "Therapod dinosaurs, such as *Allosaurus* and *Tyrannosaurus rex*, walked on two hind legs rather than on all fours." The examples give us a good idea what a therapod is.

- **Restatement.** Look at these sentences: "High mountain ranges are often built by plate collision when two plates come together. The pressure forces the land to rise along the fault line, causing a mountain range to be born." Both sentences discuss the same event—the birth of mountain ranges. But in the first sentence you are given the vocabulary phrase "plate collision." The second sentence, which says the same thing, does not use the new vocabulary phrase again.

Use these strategies in the following pages to add to your vocabulary.

© 2003 J. Weston Walch, Publisher

CARS: Science, 5–6

Using Context Clues *(continued)*

Context Clues in Action

Read the following section. How can a reader use context clues to figure out difficult words?

> ### *Cyclones*
>
> Cyclones are powerful windstorms. Types of cyclones include tornadoes and hurricanes. Unlike an ordinary windstorm, in a cyclone, the wind moves in a circular direction. Winds move around a central eye where the air is very still. Around the eye, the winds may move at hundreds of kilometers per hour.

> The passage tells me that a cyclone is a kind of a windstorm in which the winds move in a circle. I know what a hurricane is, my Aunt Jane was caught in one two years ago, and I saw the radar on television. The reading tells me that tornadoes are cyclones, too. At the center of a cyclone, there is something called an eye. According to the reading, the eye is very still, even though the winds around it are moving very fast.

Even if you had never heard of cyclones before, you would know what they are after reading this passage. Why? First, the author gives you a **definition**— "Cyclones are powerful windstorms." Second, the author gives you **examples**—tornadoes and hurricanes—to give you a mental picture of this kind of storm. The author tells you that they are unlike, or **opposite,** a typical windstorm—the type of storm you have probably experienced during a thunderstorm, for instance—because the winds move in a circular pattern. And finally, we learn that a cyclone has an eye, where the winds are still.

The word *eye* is also defined, this time by **restatement.** Look at the last two sentences. The author tells us that the winds move around the eye, which is still. Around the eye, the winds move very fast.

Using Context Clues (continued)

Application Read the following paragraph, and answer the questions that follow.

Allosaurus

The *Allosaurus*, like its later cousin, *Tyrannosaurus rex*, was a carnivore, or meat-eater. Its prey of choice were small water-loving herbivores, such as the duck-billed *Edmontosaurus*. Herbivores, unlike carnivores, eat only plants. In other words, herbivores are vegetarians. In most old picture books of dinosaurs, we see *Allosaurus* standing tall on two legs, erect. However, recent research tells us that while *Allosaurus* ran on two legs, it wasn't upright as it ran. Its tail was held out to balance its large head and chest, and it ran almost horizontally, parallel to the ground.

Without looking up any words in the dictionary, answer the following questions. Write whether you used definition, opposite, example, or restatement to figure out the answer.

1. One early cousin of *Tyrannosaurus rex* is called _____ .

   Context clue: _____

2. Another word for *meat-eater* is _____ .

   Context clue: _____

3. If you are a *vegetarian*, you are a(n) _____ .

   Context clue: _____

4. Standing tall on two legs is standing _____ .

   Context clue: _____

5. An animal that runs parallel to the ground runs _____ .

   Context clue: _____

Lesson 2
Word Parts

In science, special terms are often used. Knowing word parts can help you figure out unfamiliar words. You can add up the word parts to understand the meaning of the whole word. For example, if you came across the word "phonology," you might not know what it meant. But if you saw that it combined the root "phon," which means "sound," and "logy," which means "science," you might guess that "phonology" means something like "the science of sounds."

Prefixes are word parts that are added to the beginning of a word. Suffixes are added to the end of a word. Roots are the main part of a word. These are the parts to which you can add a prefix or suffix. Also, you might find words that are compounds. These are two root words, joined together to make a third, different word.

Common Prefixes, Suffixes, and Roots

Below are lists of common prefixes, suffixes, and roots that you might see in science. Study these lists, then answer the questions on the next page.

Prefixes		Roots		Suffixes	
endo-	inside	aqua	water	-able, -ible	able to be
exo-	outside	astro	star	-arium	place relating to
extra-	beyond	bios	life	-ate	to act on
hyper-	over	bot	plant	-er, -or	one who does
hypo-	under	chrom	color	-fy	to make
meta-	change	chron	time	-ic	related to
micro-	small	derm	skin	-ine	like
photo-	light	geo	earth	-ist	one who specializes in
pre-	before	hydro	water	-itis	disease
proto-	first	luna	moon	-ity	state of being
poly-	many	mar	ocean	-logy	science, theory
tele-	far	meter	measure	-ly	in the manner of
un-	not	morph	body	-ness	state; condition
		phon	sound, voice	-nomy	naming
		phys	nature	-osis	action, process
		scop	look	-ous	having
		sol	sun		
		terr	earth, land		
		therm	heat		
		zo	animal, life		

© 2003 J. Weston Walch, Publisher — CARS: Science, 5–6

Word Parts (continued)

Word Parts in Action

Read the following passage and see how one reader figured out some unfamiliar words.

> Our class went to the <u>aquarium</u> last week. We saw many tanks of fish, along with a touch tide pool filled with <u>shellfish</u>, crabs, sea stars, and other animals with <u>exoskeletons</u>. We learned that <u>marine</u> fish cannot live in the same tanks as freshwater fish, because the salt that marine animals need would kill <u>aquatic</u> freshwater animals.
>
> Well, I have an aquarium at home, but this paragraph is talking about a large building with many tanks. *Aqua* means water, and the suffix *–arium* means "place where," so an *aquarium* is a place where there is water. Later in the paragraph, I see the *aqua* root again. *Aquatic* means relating to water, so the animals that would die in salt water are animals that live in freshwater, I guess. Here's a prefix I've seen—*exo*—that means "outside." So animals with exoskeletons would have their skeletons on the outside of their bodies. There are a couple of compound words in this paragraph, too. *Shellfish* means a kind of marine animal in a shell? Like oysters or lobsters, maybe? And the root *mar* means "ocean," so *marine* must mean "things related to the ocean."

Application

Write the meanings of the underlined words. Use what you know of word parts.

1. A caterpillar builds a cocoon and later emerges as a butterfly. This is called <u>metamorphosis</u>.
   _____

2. Warm-blooded animals are called <u>exothermic</u>.
   _____

3. <u>Astronomers</u> use telescopes and computers to do their work.
   _____

4. <u>Sandbars</u> often disappear from view during high tide.
   _____

5. Sometimes stars can be seen during a <u>solar</u> eclipse.
   _____

Lesson 3
Word Groups

In the sciences, especially, words are found in groups. By looking for "words within words," you can figure out long new words.

For instance, the root *saur* means "lizard" and is related to dinosaurs. You might guess that words like *sauropod, Brontosaurus,* and *Stegosaurus* all have something to do with dinosaurs. The *saur* part of the word appears in all three.

Parts of Speech

You might know that some suffixes change the function, or part of speech, of a word in a sentence. For instance, the suffix *–ly* generally changes an adjective to an adverb. Here are some suffixes that change words to different functions within the sentence.

Suffixes		
Noun suffixes	**Verb suffixes**	**Adjective suffixes**
-er, -or one who -ist one who specializes in -ity, -ty quality -ment result -ness state of being -tion action	-ate to act on -fy, -ify to make -ize to cause something to be	-able able to be -al related to -ile tending to -ine relating to -ish relating to -ous having the quality of

Word Groups in Action

Read the following paragraph. Then read how one reader used words within words and word parts to figure out new words.

> Life in the rain forest is diverse. This <u>diversity</u> of life is caused by <u>environmental</u> factors such as light differences and the <u>adaptability</u> of animals and plants. <u>Biologists</u> understand that the rain-forest plants and animals <u>utilize</u> all available nutrients. This leaves the soil rather poor, but makes the diversity of life very rich.

© 2003 J. Weston Walch, Publisher

CARS: Science, 5–6

Word Groups (continued)

What does *diversity* mean? I know that "diverse" means "varied," so the *-ity* ending means it is a noun. It probably means "variety"! The next word is *environmental*. Because the *-al* ending has made this an adjective, this word probably means "related to the environment." Okay, now *biologists*. By putting the word parts together, I know that biology is the science of life, so a biologist is a specialist in life science. Finally, *utilize* is a verb, by its ending, and *util* is a form of the word "use"—so *utilize* means "to use"!

Application Read the following paragraph. Then, without using a dictionary, define the underlined words.

Early Humans

Humankind is believed to have originated in Africa. Eventually, human beings moved from Africa through Europe and to points east. American natives are probably of Asian descent. The descendants of our African ancestors can now be found everywhere on the planet.

1. originated
   definition: _____

   other words related to *originated*: _____

2. eventually
   definition: _____

   other words related to *eventually*: _____

3. Asian
   definition: _____

   other words related to *Asian*: _____

4. descendants
   definition: _____

   other words related to *descendants*: _____

PART 2
Prereading

Lesson 4: Previewing

The Reading Process

Reading for content—or understanding of the subject matter—involves a few steps. Before you read, there are steps you can take to help you understand. We call these steps prereading steps. Other steps are taken during reading. Still others happen after you have already read the passage.

In this section, we will look at things you can do—strategies—to prepare yourself to understand content. The prereading strategies are the 4 Ps: **p**reviewing, **p**redicting, drawing on **p**rior knowledge, and setting a **p**urpose.

Previewing

To preview a reading, first read the title and any subtitles. Next, skim the text, or look it over quickly. Pay particular attention to any visuals, such as graphs or photos. Read the first and last lines of each paragraph and any phrases that jump out at you.

Application

Now, preview the following paragraphs. Do not read the entire passage.

On the African Savannah

The year on the African savannah begins with rain. Rain falls in the grassland and fills the watering holes. These have been dry for many months. Rain falls into the rivers, causing them to overflow their banks. Rain falls onto the parched land and turns the golden grass green again. Rain signals to the animals of the savannah that it is time to come home.

The animals have been away in the hills, where there is still water and green grass. The first ones to return are the herds of wildebeest, zebra, and water buffalo. They bring with them babies born in the upland meadows where the animals spent the dry season. They stay together, because there is safety in numbers.

Next come the elephants and giraffes. Female elephants travel in herds that encircle young calves to protect them. Giraffe calves can run soon after birth, and they can outrun almost any predator on the savannah.

Following the prey animals come the predators. They are carnivores and need meat to survive. Lions travel together in large prides, while cheetahs are solitary hunters. Jackal families

(continued)

Previewing (continued)

On the African Savannah (continued)
return to their old dens to start new families.

Some animals live off the hunting skills of others. These animals are scavengers, and they follow the predators. Buzzards and hyenas come to finish the lions' meals.

All too soon, the short rainy season ends. The weather on the savannah becomes hot and dry. Streams dry up, followed by larger rivers. The watering holes become thick with mud, but the animals still drink from them. They cannot live without water.

Slowly, the grass fades to golden brown, and the remaining leaves on the trees become tough. Weak and sick animals die or are picked off by predators.

Finally, a day comes when there is no more water, and very little food. The prey animals begin the migration to the dry-season grazing lands. And, of course, the predators and scavengers of the savannah follow them.

When it could not be any drier, or hotter, a rumble of thunder can be heard in the distance. One year has ended, and a new year has begun. And the rains begin to fall again on the savannah.

Was your preview like the one that follows?

When I looked over the article, I saw the title was "On the African Savannah," so I think this article is going to be about the grasslands in Africa. The first paragraph talks about the year beginning with rain. Probably they are going to talk about the whole year on the savannah. Then, the beginnings of the paragraphs talk about animals who come back to the savannah during the rainy season. Each paragraph mentions a different group of animals. Then the paragraphs begin to talk about the changes during the year. How the rains end, and the water dries up. Finally, the last paragraph talks about how rain starts again.

Write three words that jumped out at you during your preview.

_____

_____

Lesson 5
Predicting

Predicting After you preview a reading, you are ready to make predictions about what the reading will reveal. What do you think the main idea of the reading is? Why did the author write this? Who is the audience for this reading? What do you think you will learn?

Application Now that you've previewed the article on the African savannah, make some predictions about what the author is trying to say here.

1. Who is the intended audience? What in the article led you to this conclusion?

   _____
   _____
   _____
   _____
   _____

2. What do you think you will learn from this article? Why?

   _____
   _____
   _____
   _____
   _____

Lesson 6
Prior Knowledge

Readers hardly ever come to a subject without some knowledge of it. Even if you think you don't know anything about the subject, you probably have some information.

You have probably seen savannah animals in a zoo or on a nature program. You may have read books about them. You might know someone who has gone on safari. All that knowledge is already in your memory.

Application Let's draw on your prior experience and prior knowledge.

1. Brainstorm. Make a list of all the words you can think of that are related to this topic. Write any word that seems to be related, no matter how silly you think it might be at first glance.

   _____

   _____

2. Next, write sentences related to the topic.

   _____

   _____

   _____

3. Finally, jot down any facts you know or any ideas you might have about the topic.

   _____

   _____

   _____

How much do you already know about this subject? More than you thought?

Lesson 7
Purpose

The Writer's Purpose There are two kinds of purpose. The first kind of purpose is how the passage is meant to be read. This is the author's purpose—why he or she wrote the piece. Is the story supposed to be entertaining? Informative? Explanatory? Or something else?

The Reader's Purpose The reader's purpose is simply why you are reading a passage. Do you want a good story? Are you interested in the subject matter?

Application

1. Write a short purpose for reading "On the African Savannah."

   _____

   _____

2. Now go ahead and read the article. Did your prereading steps prepare you for learning more about the subject? Why or why not?

   _____

   _____

This chart will help you remember the prereading steps. Eventually, you will do this in your head. For now, use the chart to help you develop good reading habits and stay on task.

4-P Chart

1. Preview	2. Predict	3. Prior Knowledge	4. Purpose
Words, graphics, captions, headings, subheadings, words that "jump out"	Based on Preview, what is this reading about?	What do I already know? Brainstorm.	Why do I want to read this? What does the author want me to get out of it?

© 2003 J. Weston Walch, Publisher

PART 3
Reading Strategies

LESSON 8
Introduction to Reading Strategies

Some reading is more difficult to understand than other reading. Science reading can be harder because you don't always have a strong base of prior knowledge. On the next few pages, you will find some methods, or strategies, for making this reading easier to understand.

Some strategies are research-oriented. They help you organize your thoughts in order to write a report or prepare for a quiz. Other strategies help you remember more about the reading. All of these strategies involve writing, because writing about any subject helps to solidify knowledge.

Graphic Organizers

There is an old saying that pictures are worth a thousand words. Often, seeing something visually helps us remember more than just reading. We can organize our reading in a visual way with a **graphic organizer.** Many graphic organizers are used in science. For instance, here is a simple kind of graphic organizer, called a **cycle:**

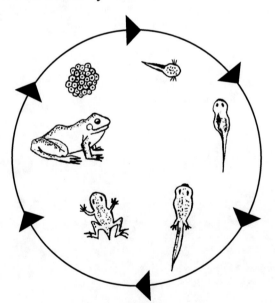

The Frog's Life Cycle

Graphic organizers give us a lot of information at a glance. They can be a good way to show complex information. In this book, we will learn five different kinds of reading strategies that use graphic organizers.

Introduction to Reading Strategies *(continued)*

- **KWL (What I KNOW, What I WANT to know, What I LEARNED).** In this technique, you list things you already know, what you hope to learn, and then, after the reading, what you actually learned.
- **SQ3R (Survey, Question, Read, Retell, Review).** In this strategy, you pose questions to yourself, then answer them after reading the selection.
- **Semantic Web.** You can use this approach to categorize your reading and organize categories into a graphic device.
- **Outline.** This technique helps you make sense of what you read by noting the main and secondary points in any reading passage.
- **Structured Notes.** This process helps you take notes more efficiently, whether you are taking notes while reading or while listening to someone talk.

1. Think about the kinds of reading you do every day. List some of the reading you do on a daily basis.

   _____

   _____

   _____

2. From that list, think about when you read long passages for information. When do you do this type of reading?

   _____

   _____

Lesson 9
KWL

The KWL strategy is a good technique to help you use your prior knowledge.

K stands for using what you **KNOW** about the subject.
W stands for determining what you **WANT** to know.
L stands for recording what you **LEARN** while reading.

As you can probably guess, you will use some of your prereading skills with this approach. To use the KWL strategy, you first have to preview the article.

KWL in Action

Make a chart with three sections. The first section will be all the things you know about a subject. The second section will list all the things you want to learn from the reading. The third section, which you will fill in after you have read the passage, will list the things you have learned about the subject.

Let's say you have to read the following paragraphs about butterflies. First, you would fill in the **K** column. Then you would preview the article and fill in the **W** section.

Life of a Butterfly

You probably already know that butterflies don't hatch, with wings, from an egg. Instead, the larval—or juvenile—form of a butterfly is called a *caterpillar*. Caterpillars have as many different colors and features as adult butterflies do!

The mother butterfly lays her eggs on a very specific kind of leaf. These are the only leaves that the caterpillar will eat when it hatches. And caterpillars live to eat. They munch away 90 percent of the time they are awake.

Eventually, caterpillars spin a small enclosure called a *chrysalis*. A chrysalis is like a cocoon. But only caterpillars that will become moths spin cocoons.

For several weeks, the chrysalis does not change. This is a dormant period for the organism. Finally, however, a butterfly emerges. It takes about a day for its wings to dry completely. Butterflies, unlike caterpillars, drink the nectar of flowers. They live to mate and lay eggs. And then the cycle begins again.

KWL (continued)

K What I KNOW	W What I WANT to Know	L What I LEARNED
• Caterpillars are baby butterflies. • Caterpillars change into butterflies.	• How do caterpillars change? • What do caterpillars eat?	• Caterpillars eat the leaves they were born on. • Butterfly cocoons are chrysalises.

> I do know something about butterflies. They begin as caterpillars and change into butterflies later. The title is very clear—it talks about the life of a butterfly. How do caterpillars change into butterflies? And what do they eat when they are still caterpillars?

After reading the article, you fill in the last section.

> I learned that caterpillars will only eat the leaves they hatched on. I also learned that the name for a butterfly cocoon is a "chrysalis."

Application Follow the prereading steps for the article "Forest for the Trees." Fill out the **K** and **W** sections on the chart that follows. Then read the article, and fill in the **L** column.

Forest for the Trees

When you think of a forest, what do you mainly think about? Trees? Most people do. Trees are an important part of a forest. But they are not the only part.

Trees are plants. While they are the largest and most impressive plants, they are not the only plants that grow in a forest. Take a walk in any forest or wooded area. You will find small wildflowers, weeds, and small shrubs covering spaces between the trees.

On fallen trees and damp ground, you will also see many different kinds of fungi. Mushrooms, lichens, morels, and truffles all live on dead and rotting vegetable and animal matter. They help decompose organic matter, breaking it down into smaller parts. Without fungi, the forest would soon be a vast pile of deadwood and leaf litter.

And, of course, in any forest, there are thousands of species

(continued)

KWL (continued)

Forest for the Trees (continued)

of animals. Everything from the tiniest insect to the largest mammal plays an important role in the life of the forest. Insects fertilize the plants. Spiders and birds keep insect life in check. Birds also play a role in dispersing seeds. Large mammals keep the forest healthy by fertilizing it.

If you visit a forest and look only at the trees, you are missing most of what a forest is.

K What I KNOW	W What I WANT to Know	L What I LEARNED

KWL (continued)

QUIZ: Forest for the Trees

Answer the following questions. In 1, circle the letter of the correct answer. In 2–5, write your answer in the space provided.

1. Which definition describes a forest?
    (a) a lot of trees
    (b) trees and animals
    (c) all living things in the area
    (d) deadwood

2. According to the article, what kinds of plants other than trees can be found in a forest?

    _____

3. What is the importance of fungi in the forest?

    _____

4. To what kingdom of living things do trees belong?

    _____

5. Imagine a forest without animal life. How would it be different? Why?

    _____

    _____

    _____

Lesson 10
SQ3R

SQ3R stands for Survey, Question, Read, Recall, and Reflect. This strategy can help you focus your thoughts before reading a passage, search for new information, and remember what you read. To organize your ideas, you can use a graphic organizer with one box for each step.

The first step is to **survey.** This is another word for "preview." Scan the text quickly. Record anything you think might be important in the "survey" box.

Next, ask yourself **questions** about the article before you begin to read. You might have questions about the subject, the author's point of view, or your previous knowledge in the area. Write these questions in the "question" box.

Now, **read** the article. While you read, look for the answers to your questions. After you read, record the answers you find in the "read" box.

Next, **recall** what you have learned. Using your own words, quickly retell what you've just read. Write that in the "recall" box.

The last step is to **reflect** on what you learned. Check to see if your questions were answered fully. Write any new ideas in the "reflect" box.

SQ3R in Action

We will try it once together with the following short passage. Read the passage. Then read the paragraph that shows one reader's thoughts about the article.

How Birds Fly

Did you ever wish you could fly? Humans would find it very hard to get off the ground, even with wings. This is because, unlike birds, we have solid bones. Birds have hollow bones that are very light. However, the most important reason birds can fly has to do with the shape of their wings. Bird wings are designed a bit like airplane wings. On the top of the wing, air has to travel over a small hump. This makes the air flow faster than at the bottom of the wing. When there is a difference in air flow, there is also a difference in air pressure. The top of the wing experiences a lower air pressure than the bottom, giving the bird lift. Even with this lift, birds have to flap their wings very fast to stay aloft unless they are floating over an updraft, called a *thermal*. Birds who ride the thermals don't have to flap their wings at all.

SQ3R (continued)

> This paragraph is about how birds fly. From scanning the reading, it seems like birds have something called "lift" because of the shape of their wings. They also have hollow bones, which makes them light. Some birds float on top of a rise of air called a *thermal*.

S Survey	Q Question	R Read	R Recall	R Reflect
• Birds have hollow bones. • Birds are very light. • Birds have lift because of the shape of their wings. • Some birds float on thermals that rise from the ground.	• How does the wing shape help? • Why do hawks and eagles look like they are soaring?	• There is a difference in the air flow above the wing and under it, causing an air pressure difference. • They float on air that rises from the ground and basically float in the sky.	• Birds can fly because their wings are shaped so that there is an air pressure difference above and below.	• Light bones • Wings like airplanes

Write the main points of the reading in the "read" column, and repeat them in your own words in the "recall" column. The act of writing something in a few different ways helps you remember it longer.

The "reflect" step is done mostly in your head, or with a friend or teacher. Review the questions you asked yourself. Did the reading answer the questions for you? If not, where can you go for more information?

SQ3R (continued)

Application Survey the article "What Is a Constellation?" Fill in the first column of the SQ3R chart that follows. Then read the article and complete the chart.

What Is a Constellation?

When you look into the sky on a clear night, you see about 2,000 stars. Some of them you can see with your naked eyes, and some you can only see with telescopes. They are grouped into 88 different star groups, or constellations.

Constellations are supposed to represent dot-to-dot drawings of everything from people, to animals, to scientific instruments. They are a line-of-sight effect. That means that, although we can see the stars in a group, they might be very far away from one another in space. Why?

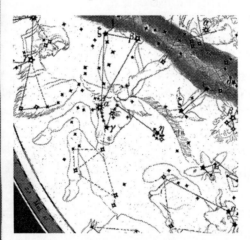

The sky isn't a bowl or a flat sheet of paper. The universe also has depth. Some stars in a constellation might be far behind others. To us, they look as if they are right beside each other. Still, in space, they might be millions of kilometers apart.

Some constellations seem to mark the path of the sun and the moon. These constellations are called the *zodiac*. When the sun is in a particular constellation, children born at that time are said to share that zodiac sign. For instance, the sun is in the constellation Aquarius during late January and February. So, people born during late January and February are Aquarians. There are 12 zodiac constellations, and it takes the sun 12 months to go through all of them.

Many of the constellations are related to Greek mythology. Hercules, for instance, is a bright autumn constellation. So is Pegasus, which is supposed to be a beautiful winged horse but really looks like a square with stars sticking out of it. It takes a lot of imagination to see why some constellations are named the way they are!

SQ3R *(continued)*

Now that you've read the article about constellations, fill in the remainder of the chart.

S Survey	Q Question	R Read	R Recall	R Reflect

SQ3R (continued)

QUIZ: What Is a Constellation?

Answer the following. In 1–2, circle the letter of the correct answer. In 3–5, write your answer in the space provided.

1. The word *constellation* means
   (a) a group of stars.
   (b) a picture.
   (c) a solar system.
   (d) Hercules.

2. Constellations are a line-of-sight effect. This means
   (a) they represent dot-to-dot drawings.
   (b) they are far away.
   (c) they are together in space.
   (d) they only look like the constellation from Earth.

3. When is the sun in the constellation Aquarius?
   _____

4. How long does it take the sun to move through all 12 zodiac constellations?
   _____

5. Why do you think most constellations are named for Greek mythology?
   _____
   _____
   _____
   _____

Lesson 11
Semantic Web

Semantic means "relating to words." A semantic web is a way to look at words and ideas as you read a passage.

To use the strategy, choose a word or an idea from the reading. Write this word or idea in the center of a sheet of paper. Next, brainstorm a list of words or ideas that are related to your first word. Group the words and concepts you have written into categories. Then link them to the central word with a "web."

Semantic Web in Action

Let's do one together. Read the following short passage.

Bald Eagles

Not long ago bald eagles were endangered—at risk of becoming extinct. Special measures were taken to protect them. One of these was outlawing a pesticide, or bug killer, called DDT. DDT caused eagles to lay eggs with very thin shells. The mother eagle would crush her own eggs by sitting on them. Since DDT was banned, eagles have made a strong recovery.

Eagles usually live in remote areas. However, this is not their only *habitat*, or place where they live. They may also be found near clean rivers in small towns. They are raptors, or birds of prey. This means they hunt and eat meat. Their diet is fish, small mammals, and small reptiles.

Eagles build their nests in tall trees, or sometimes at the top of power poles. The nests are very large and are called *aeries.*

Usually, a female eagle lays one to three eggs. The eggs hatch about 35 days later. When the eaglets hatch, they are helpless at first. Both parents feed the young with shredded pieces of meat. The eaglets grow very quickly. When the eagles are three months old, they are able to fly.

You can see that the major idea in this passage is the bald eagle. Those will be the two words you place at the center of your semantic web. Now, brainstorm a list of things you know about eagles, either from your prior knowledge or from the reading.

Semantic Web (continued)

> I know that bald eagles live in remote regions. I remember seeing one during our vacation in Alaska! They are birds of prey, or raptors. I think they are related to golden eagles, falcons, and hawks. I saw a nature program about how eagles are being reintroduced to places where they did not survive. That means that eagle habitat is increasing. The article says that a mother eagle lays one to three eggs, and that both parents feed the young. The article also talks about how eagles almost went extinct because of this pesticide, DDT.

Now we have a lot of information. The next step is to categorize it. The first thing mentioned here is where the eagle lives—its habitat. Write the word "habitat." Now draw a line to connect the words "bald eagle" and "habitat." What do we know about the eagle's habitat? We know eagles live in remote areas and near clean rivers in small towns. We also know that the eagle's habitat is increasing and that eagles are being reintroduced to other areas. All of those things relate to "habitat." Write each piece of information in its own circle. Draw lines to connect related information.

What other categories can we use here? The passage talks about what eagles eat. We know eagles are related to other kinds of birds, and that they teach their eaglets to kill prey. We also know that they nearly became extinct because of a pesticide, DDT. We can add all of this to the web.

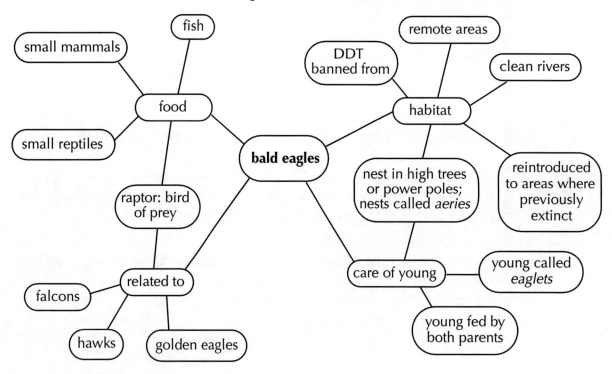

Semantic Web (continued)

Application Read the following article. Fill in the semantic web before, during, and after reading. Don't forget to add your own personal knowledge to the web. Each web is different. Add lines and circles where needed.

Salmon Run

Every summer in Juneau, Alaska, the salmon come home. In shallow streams, the pink-fleshed fish perform the last act of their lives—spawning the next generation. In a few days, the bodies of those not eaten by bears and scavengers will wash down the streams, into rivers, and at last into the sea.

The fish die in the same streams in which they were born. It is only a lucky few that make it this far. Most salmon eggs never even hatch. They are eaten by other fish and amphibians. Young salmon, called *fry,* live in the protected creeks for several months after hatching. However, most of the fry that hatch are soon eaten by predatory birds and small mammals. When the fry are slightly older, they move into larger bodies of water—lakes, ponds, or slow-moving rivers. After a year, they begin their migration, or move, to the sea.

The journey is not easy. Many rivers have been blocked by dams. There are predators everywhere. A lucky few make it to the ocean. They spend most of their adult lives there.

There is no safety for the salmon in the sea, either. They are prey to many larger fish and are fished for by people who love the sweet taste of the fish. Pollution in the water kills some salmon.

However, some do survive. And, at last, they return to the streams where they were born. They swim upstream, against the current, over fish ladders made to help them get over dams, in order to keep the species going for another generation. After a mating ritual, the male builds a nest. The female lays thousands of eggs, which are fertilized by her mate. The parents guard the nest as long as they can. Then they die. Their babies will be born in the spring, and the cycle continues.

Semantic Web (continued)

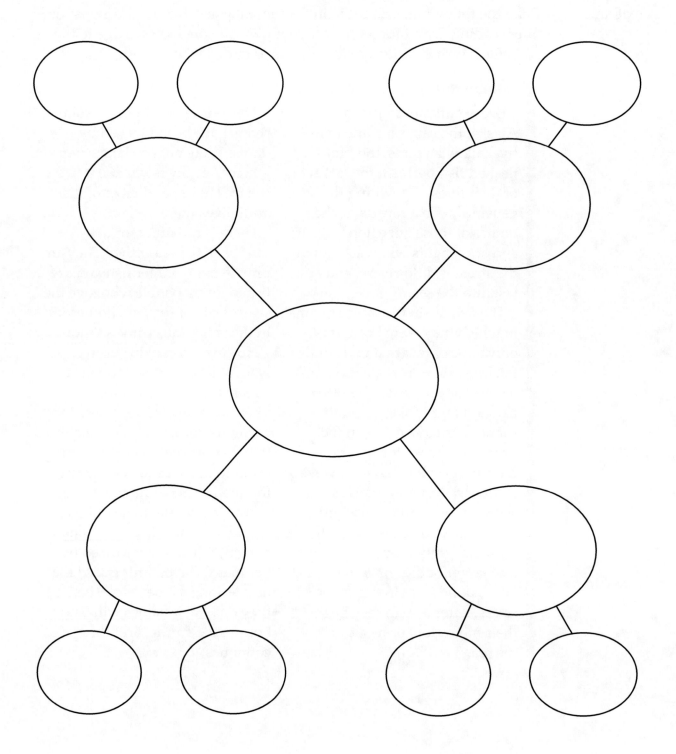

Semantic Web (continued)

QUIZ: Salmon Run

Answer the following. In 1–2, circle the letter of the correct answer. In 3–4, write your answer in the space provided.

1. In the passage about salmon, the word *spawning* probably means
   (a) producing young.
   (b) swimming upstream.
   (c) feeding.
   (d) dying.

2. The word *predatory* is related to two other words later in the passage. These are
   (a) protected and fry.
   (b) predator and protected.
   (c) predator and prey.
   (d) prey and protected.

3. What is the greatest threat to salmon completing their life cycle?

   _____
   _____
   _____

4. How have dams and other human activities affected the salmon's life cycle?

   _____
   _____
   _____
   _____

Lesson 12
Outline

Outline

You may already be familiar with making an outline to prepare for writing an essay or a paper. This technique can also help you organize and make sense of what you read.

First, preview the passage, then make an outline. The most common method is to use Roman numerals and the letters of the alphabet to structure the outline. The Roman numerals show major ideas; capital letters show secondary ideas. If necessary, break the outline into more ideas by using Arabic numbers (1, 2, 3 . . .) and lowercase letters.

Usually, an author groups a major idea and its supporting ideas into one or two paragraphs. You can use this as a general guide. However, there may be things earlier or later in the passage that you will want to outline.

Outline in Action

Let's try one together. Preread to find the main points of the reading. Then read the article and follow the outlining process of one reader.

American Alligators

American alligators live in warm, swampy areas of the Deep South of the United States. They are most common in Florida and Louisiana, but can also be found in Mississippi, Alabama, and Georgia. They prefer fresh or mildly salty swampland. The Florida Everglades, a national park in southern Florida, is a perfect home for many alligators.

Alligators are reptiles. They are cold-blooded, so they must spend hours every day warming in the sun. Their favorite foods are fish and amphibians. However, they will also eat mammals, birds, and other reptiles.

American alligators were seriously endangered a few years ago because their home territories were being drained for new houses and they were being hunted for sport. However, they have made a remarkable comeback. Habitat depletion is still a problem. Because of this, alligators are often found in human areas.

Outline (continued)

> This article is about alligators. The first paragraph talks about their habitat. The second paragraph talks about the animal itself, that alligators are reptiles, how they spend their days, and what they eat. The third paragraph talks about how they were endangered a few years ago but have now recovered, although they are now living in human areas, which is a problem.

I. Alligator habitat
   A. swamps
   B. Deep South
      1. Florida
         a. Florida Everglades
      2. Louisiana
      3. Alabama
      4. Mississippi
      5. Georgia

II. The alligator
   A. reptile
      1. cold-blooded
      2. spends a lot of time warming in sun
   B. eats
      1. fish
      2. amphibians
      3. small mammals
      4. reptiles

III. Alligator problems
   A. territories lost due to new home construction
   B. once seriously endangered
   C. sometimes found in human areas

Outline (continued)

Application Apply your prereading skills to the following reading. Figure out main ideas for your outline. Then read the article and fill in the minor ideas in your outline.

What Is a Solar Eclipse?

The air cools, the birds quiet down, and the crickets start chirping. The sky darkens. Slowly, the sun is replaced by darkness. The end of an ordinary summer day? No! This happens at 1:00 in the afternoon! One lucky region in the world is experiencing a solar eclipse.

Solar Eclipse, 1999

What is a solar eclipse? A solar eclipse occurs when the moon is in a direct path between the earth and the sun. (This can only happen at the lunar phase called the *new moon*.) For viewers on Earth, the moon will block the sun from sight. During a total eclipse, the entire bright face of the sun is covered by the moon. This will only be seen by people in a narrow band of the earth. Most of the world will either notice nothing or will have a partial eclipse, where the moon covers part of the sun's disk.

As the moon moves across the face of the sun, people who have armed themselves with welder's glass can watch the spectacle. The moon eats out a portion of the sun, until at last the moon covers the entire face of the sun. Only the bright halo of the sun, called the corona, can be seen during totality. Totality doesn't last long. Slowly, the moon moves away and the sun shines again. The birds sing again, and the insects quiet down. The air loses its chill. It is early afternoon again.

Never look at the sun, even in total eclipse, with the naked eye. It can cause serious eye damage. Instead, get a piece of strong welder's glass, or make a pinhole viewer, before a solar eclipse.

Although partial eclipses occur regularly, a total solar eclipse is a rare and spectacular sight. Don't miss it!

Outline *(continued)*

"What Is a Solar Eclipse?" Outline

I. _____

    A. _____

    B. _____

    C. _____

    D. _____

II. _____

    A. _____

    B. _____

    C. _____

    D. _____

    E. _____

    F. _____

    G. _____

III. _____

    A. _____

    B. _____

Outline (continued)

QUIZ What Is a Solar Eclipse?

Answer the following. In 1–2, circle the letter of the correct answer. In 3–5, write your answer in the space provided.

1. In the passage you just read, the phrase *total eclipse* probably means
    (a) the sun entirely covers the moon.
    (b) the moon entirely covers the sun.
    (c) the sun has set.
    (d) it is night.

2. Solar eclipses are possible only during the
    (a) full moon.
    (b) first quarter.
    (c) crescent moon.
    (d) new moon.

3. What are three things that happen when the sun goes into eclipse?

    _____

    _____

4. Why do birds stop singing at the beginning of a solar eclipse?

    _____

    _____

5. Why do you think only certain areas on the earth can see a total eclipse?

    _____

    _____

    _____

LESSON 13
Structured Notes

In this section, we will talk about how to read and take notes so that you get as much out of reading as possible. To use structured notes, you must figure out the central idea and the way the reading is organized. Then you can make a graphic organizer to keep the arrangement clear in your mind.

Structured Notes in Action

Let's do one together. First, preread to get a good idea of how the article is arranged. Then read it and follow the structured notetaking of one reader.

Momentum

Everything has mass—the amount of matter an object contains. We usually measure mass in kilograms. If the object is very small, we measure mass in grams. Mass is pulled on by Earth's gravity. Because of this pull, we feel mass as weight. So on Earth, our mass is equal to our weight, although that would not be the case in space!

Sometimes, mass is in motion. Any time motion is in a particular direction, it is called *velocity*. You already know that falling off your bicycle at a dead stop hurts less than falling off when you are riding at 10 km/hr. This is because your mass—you and the bicycle—is moving with velocity, and that gives you momentum. To find momentum, we multiply mass times velocity. If you have a mass of 20 kg and you are riding along at 10 km/hr, your momentum would be 200 kg km/hr. If an object is not in motion, there is no momentum, even if the object is massive.

The author is talking about three different words in physical science: mass, velocity, and momentum. To understand momentum, we first have to understand mass and velocity. Then the author gives us an example.

Structured Notes (continued)

Let's construct a graphic organizer.

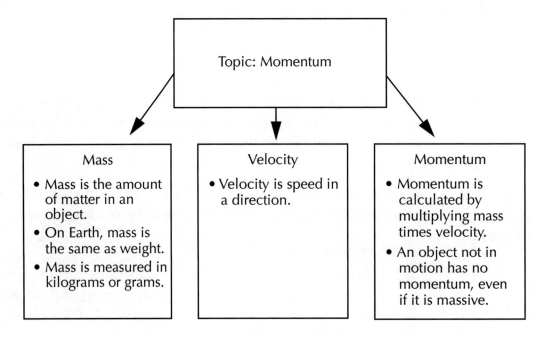

Application Use your prereading skills to identify categories for structured notes. Then read the article, taking notes on the graphic organizer.

The Remarkable Insect Body Plan

There are thousands of species of insect. Billions of individual insects, of every size, color, and shape imaginable, live in nearly every climate on Earth. They all look very different from one another. Bees do not look like ladybugs. Ants do not look much like butterflies. Houseflies do not look like crickets. Yet they all have one fundamental thing in common. They all have the same body plan.

The insect body plan is a remarkably simple and versatile one. It is based on three segments—the head, the thorax, and the abdomen.

On every insect, the head contains antennae, eyes, and mandibles. Mandibles are the insect's jaws and mouthparts.

(continued)

Structured Notes (continued)

The Remarkable Insect Body Plan (continued)

The details of the parts of the head vary from insect to insect.

On every insect, the thorax is the body segment to which the legs and wings are attached. It is the part in the middle of the insect. Wingless insects, such as ants and termites, still have places where the wings used to attach.

The abdomen of every insect contains most of the vital organs. Although insects vary wildly in appearance, their body plan is exactly the same.

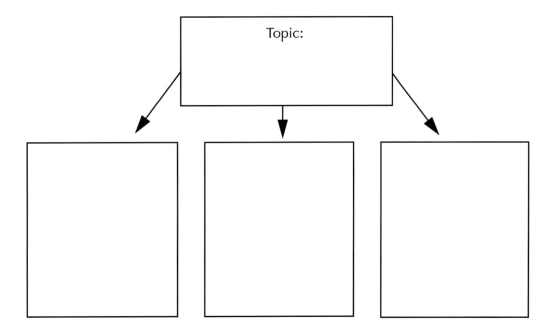

Structured Notes (continued)

QUIZ: The Remarkable Insect Body Plan

Answer the following questions. In 1, circle the letter of the correct answer. In 2–5, write your answer in the space provided.

1. In the previous passage, what does the word *mandibles* mean?
   (a) hands
   (b) eyes
   (c) jaws
   (d) antennae

2. How many body segments does an insect have? Are there any exceptions?
   _____
   _____

3. What important things can be found attached to the thorax?
   _____

4. What special things are found on the head of an insect?
   _____

5. Why do you think insects evolved with such diversity? How did the body plan assist in this variety?
   _____
   _____
   _____
   _____

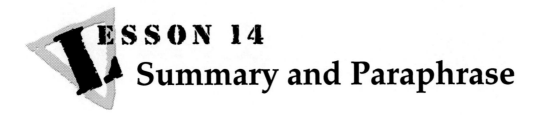
Lesson 14
Summary and Paraphrase

Now you know some of the strategies you can use before and during reading. This lesson will give you a strategy you can use after reading.

- **Summarizing.** A summary briefly recounts the reading, using the author's words.
- **Paraphrasing.** Paraphrasing retells the main points of the reading, using your own words.

Some of our reading strategies used summarizing and paraphrasing. For example, the SQ3R technique had a built-in "recall" component. The **L** column in KWL charts asks what you learned from a reading.

Summarizing and Paraphrasing in Action

Let's try one together. Read the following article. Then we'll look at the difference between a summary and a paraphrase of the same selection.

Seeking Its Own Level

When you pour water from a pitcher into a glass, what do you notice about the water levels? Does the water "stick" to the sides of the pitcher, or stay at the top of the glass? Of course not. Water "wants" to go down as far as it can. Water seeks its own level. Ultimately, that level is sea level.

On a large scale, water starts a journey high in the mountains. Each little spring or stream flows downhill to join up with larger rivers. These small streams are called *tributaries.*

The larger rivers also flow downhill. Sometimes it may not look that way. Rivers can look sluggish; they may seem not to move at all. However, the river is running in the right direction to get its water to the sea. The only exception to this rule is close to the sea, where the tides may push water back up through the river channel at high tide. These kinds of rivers are called *tidal sloughs.*

© 2003 J. Weston Walch, Publisher

Summary and Paraphrase (continued)

Let's look at a summary, which uses words from the reading, and a paraphrase, which uses your own words.

Summary

> Water poured from a pitcher to a cup settles down in the cup and the pitcher because water seeks its own level. Water is trying to get down to sea level. In mountains, water always runs downhill, whether the water comes from a tributary or a big river. Rivers always run in the direction they need to run to get to the sea. The only exception is at high tide, when water may back up into river channels. These are called *tidal sloughs.*

Paraphrase

> When I pour water, the water goes to the bottom of whatever container I pour it into, because water "wants" to be as low as it can be. That's called seeking its level. It really wants to be at sea level. Streams and rivers always run downhill and toward the ocean, except at high tide, when some water backs up along river channels.

A summary is a quick review of the article. As you can see, the language is almost the same. Vocabulary that was introduced in the article is seen again in the summary. New concepts are repeated, often without the explanation found in the full article.

In the paraphrase, however, you take the time to write it in such a way that you will understand it later. You might change some of the author's terminology so that you understand the general idea.

Both methods help you gain a better understanding of what you have read.

Summary and Paraphrase (continued)

Application — Now, read the article "Volcanic Rock," then write a summary and a paraphrase.

Volcanic Rock

There are three main kinds of rock on the planet. They are called *igneous*, *sedimentary*, and *metamorphic*. Most of Earth's crust is made up of igneous, or volcanic, rock. Volcanic rock is formed within volcanoes.

There are several different types of volcanic rock. Each is different from the others in important ways. The most common form is basalt. Basalt can be gray, black, or pink. It is usually flecked with small air pockets or holes. Under pressure, basalt will eventually become granite. Occasionally, basalt forms hollow, ball-like structures called *geodes*. Over time, crystals may form in these structures.

An interesting kind of rock is pumice. Pumice is a light rock, filled with air pockets. Some pumice is so light that it can float in water. You may have seen a pumice stone; some people use them for pedicures. Pumice is usually light gray or even white.

Obsidian is sometimes called "volcanic glass." Unlike basalt and pumice, it is heavy and dense, and contains little or no air. Obsidian can be sharpened to fine points. People have used it to make arrowheads and knife blades for centuries. Obsidian is usually black and shiny, but pink obsidian can occasionally be found.

Geodes

Summary and Paraphrase (continued)

Summary

_____
_____
_____
_____
_____
_____
_____
_____
_____
_____

Paraphrase

_____
_____
_____
_____
_____
_____
_____
_____
_____
_____

PART 5
Reading in Science

Lesson 15
Common Features and Patterns in Science Reading

Where do you come across readings in science? In school, you may read them in textbooks and workbooks. You may read newspaper and magazine articles about new advances in science. In daily life, you may read about a coming storm on the Internet. You may have to make sense of a medication insert, or a mailing from your town about water quality in your community.

What sorts of reading in science do you do on a regular basis? List a few of them below.

_____

_____

How are scientific readings different from other general readings? A scientific reading may contain certain features and patterns, such as these.

- **Special Terms.** These are words that are specific to the reading. In order to understand the topic, you have to understand the words; therefore, many scientific readings contain built-in definitions or examples.
- **Topic and Subtopic.** Often, a scientific reading takes a large idea and breaks it down into logical subtopics.
- **Classification.** A classification is a type of organizational technique that helps sort objects from a broad group into smaller groups.
- **Steps in a Process.** Often, scientific readings are how-to manuals for repeating experiments or making equipment.
- **Assertion and Support.** Much of science hinges on showing how an idea (an assertion) is supported by evidence.

It is important to recognize these patterns as you read scientific passages. Knowing the pattern will help you figure out which reading strategy will help you most.

Lesson 16
Special Terms

Science Reading

The goal of scientific reading is to communicate new ideas. So, you can expect that new terminology will be used. Sometimes the author defines a new term for you. However, this isn't always the case. In a school textbook, or a newspaper or popular science magazine, authors usually define anything they think readers might not know. However, in scientific journals, which are written for people who study and work in the sciences, the reader is expected to understand the terminology. Only new words will be defined.

Even without a definition, there are things you can do to understand new words. Use the vocabulary strategies from Part 1 of this book:

- Pick up context clues.
- Keep in mind word parts.
- Remember that science terms are often built upon one another.

These strategies will help you get more out of your reading while building your scientific vocabulary.

Application

Now read this passage, and answer the questions that follow it.

The Age of Mammals

If the Cretaceous extinction had not occurred, the world might still be populated by dinosaurs or their large <u>reptilian</u> descendants. Mammals today might well live their lives as scavengers and <u>insectivores</u>.

As it happened, however, the small mammals at the beginning of the Cenozoic Era about 65 million years ago began a rapid <u>diversification</u> in both <u>herbivores</u> and <u>carnivores</u>. Soon after, <u>cetaceans</u> returned to the seas and bats took to the air.

There was a general increase in size in some species, but most mammals remained quite small. Over half the 4,000 living species of mammals today are <u>diminutive</u> rodents. When dinosaurs were alive, the majority of <u>terrestrial</u> animals were larger than the modern cow; today, the reverse is true.

Special Terms (continued)

In the reading about mammals, several words were underlined. The following questions relate to some of those words.

1. The word *reptilian* probably means

    (a) like a dinosaur.   (b) like a fossil.   (c) like a reptile.   (d) like a mammal.

    What strategy did you use to determine this? _____

2. The word *insectivores* probably means

    (a) scavengers.   (b) meat eaters.   (c) frogs.   (d) insect eaters.

    What strategy did you use to determine this? _____

3. The word *diversification* probably means

    (a) stocks and bonds.   (b) change over time.   (c) becoming different kinds.   (d) underwater.

    What strategy did you use to determine this? _____

4. The word *cetaceans* probably means

    (a) ocean mammals.   (b) lobsters.   (c) crabs and shrimp.   (d) insects.

    What strategy did you use to determine this? _____

5. The word *diminutive* probably means

    (a) small.   (b) gigantic.   (c) fierce.   (d) ocean-dwelling.

    What strategy did you use to determine this? _____

6. The word *terrestrial* probably means

    (a) earthlike.   (b) ocean-dwelling.   (c) land-dwelling.   (d) airborne.

    What strategy did you use to determine this? _____

Lesson 17
Topic and Subtopic

Topics and subtopics are used in most expository (explanatory) writing. Scientific writing is only one example. In science, the way the author breaks down subtopics reveals the *thesis statement*—what the author hopes to prove or show to the reader. Any big topic can be broken down in any number of ways. An article about modern astronomy might be broken down by subgroup—radio astronomy versus visual astronomy, for instance. Or the author might want to point out the differences between American astronomy and Swiss astronomy, or the importance of the Hubble Space Telescope in new astronomy. The way an author uses topics and subtopics gives us a clue about his or her biases and what we are being asked to "get" from the writing.

How do we recognize a subtopic while reading? Sometimes, although not always, a subtopic is "announced" by a subhead. Subtopics are always set apart in their own paragraphs. However, one subtopic might cover several paragraphs. Watch for a verbal change of direction.

Application   Read the following short passage. Keep your "critical eyes" open for how the author uses subtopics to further a thesis. Then answer the questions that follow.

> **Life on Early Earth**
> Earth was formed 4.6 billion years ago. The earliest fossils ever found date from a billion and a half years later—3.1 billion years ago. What was life like when our planet was young?
>
> **Bacterial Soup**
> Early life on Earth existed solely in the oceans. Undoubtedly, the earliest life-forms on Earth were bacteria. The oldest bacteria, archaebacteria, could not handle free oxygen in the water or in the atmosphere. They were chemical processors. They had no real need for sunlight, either. Today, many species of archaebacteria still live near thermal vents and in sulfur springs.
>
> However, a new species of bacteria arose—cyanobacteria. This species used sunlight; they were photo processors. They also gave off a deadly waste product
>
> *(continued)*

Topic and Subtopic (continued)

Life on Early Earth (continued)

that polluted the earth's early oceans and atmosphere—oxygen. Many of the older species of bacteria died out completely. But the oxygen made by the cyanobacteria made it possible for other forms of life to arise.

Old Softies

The next on the scene were single-cell and multicellular animals with soft bodies. Paramecia, algae, and other single-cell organisms took over the oceans, followed by large, soft-bodied creatures such as jellyfish.

1. What is the main topic of the reading?
   _____

2. What are the subtopics? How can you tell? Would you have been able to recognize the shift in topic without the subheads?
   _____
   _____
   _____

3. How is the author using subtopics to lead us to the thesis statement?
   _____
   _____
   _____

4. Based on the way the passage is organized, what do you think the author wants us to learn from it?
   _____
   _____
   _____

5. Was the topic/subtopic a good organization for this subject? Why or why not?
   _____
   _____
   _____

Lesson 18
Classification

Classification is widely used in science writing. Scientists group many things in terms of what they have in common—their shared characteristics. In fact, a lot of science is making these comparisons.

One example you may already be familiar with is the classification system for life. This is known as the *Linnean Taxonomy* (science of classification). Here's an easy way to remember the system's order: King (Kingdom) Philip (Phylum) Came (Class) Out (Order) From (Family) Germany (Genus) Singing (Species).

Every living thing—animal, plant, bacterium, fungus, or protoctist—can be organized using this system. The system starts out with the largest group of organisms, then narrows down to smaller groups, until only one species remains. Here is our human family tree.

Kingdom:	Animalia
Phylum:	Chordata
Class:	Mammalia
Order:	Primates
Family:	Hominidae
Genus:	*Homo*
Species:	*sapiens*

Most of the time, only the last two classifications—genus and species—are used to identify an animal or a plant. We certainly don't identify humans using all these words! *Homo sapiens* will do fine. However, all these classifications apply, and from time to time they may be very important. It may be important to note that humans are also primates, for instance, or that humans are also animals.

Classification *(continued)*

Application   Read the following short passage. Then answer the questions that follow.

> *Diatom*
>
> Are you aware that the animal and plant kingdoms are not the largest ones on the planet? The largest kingdom is the kingdom of the protoctists.
>
> Protoctists include red, brown, and green algae, as well as most of the single-cell organisms that make up life in the average pond. Amoebae, dinoflagellates, and paramecia are all members of this large and varied kingdom.
>
> One of the most interesting and beautiful phyla of the kingdom protoctista are the Bacillariophyta, or the lovely diatoms. There may be 10,000 or more species. There are two classes of diatoms, the Centrales and the Pennales. The Centrales are radially symmetrical, like a sea star. The Pennales are bilaterally symmetrical, like a clam. One Centrales order, the Thalassiosira, form great colonies. One particular species, *Thalassiosira nordenskjoldii*, lives under the Arctic icecaps and forms large colonies for mutual protection.

1. Create as much of a family tree as you can for the diatom *Thalassiosira nordenskjoldii*. Note that the order name and the genus name are the same. One space will be blank.

   Kingdom: _____

   Phylum: _____

   Class: _____

   Order: _____

   Family: _____

   Genus: _____

   Species: _____

2. The author talks about two classes of diatoms. How do they differ from each other?

   _____
   _____
   _____

Steps in a Process

Much science writing describes how research or experiments were done. Because of this, some science writing looks like a cookbook or how-to manual. The reason for this is the scientific method.

The scientific method, which we use to this day, was first described in 1637. It sets up procedures for experimentation. These include identifying a problem to be solved, a hypothesis (an educated guess) about the problem, a list of materials to be used, a procedure for the test, the data for the test, and then, finally, the conclusion—whether or not the hypothesis was correct. The last part of the scientific method is getting one's results replicated, or repeated.

Application Read the following passage. Then answer the questions that follow.

Velocity Experiment
Problem: Finding the speed of an object
Hypothesis: Velocity is a change in distance over time.
Materials list: Small battery-driven car, two metersticks, stopwatch, little bits of paper
Procedure:
1. Lay the metersticks end to end.
2. Starting at the end of the first meterstick, release the car and start the stopwatch.
3. Every two seconds, drop one bit of paper where the car's front wheels are.
4. Record the position of the car at two-second intervals, remembering to add the additional 100 cm for each meterstick.
5. Graph the change in distance over the change in time. (See graph on next page.)
6. Divide distance by time to arrive at the velocity.

(continued)

Steps in a Process (continued)

Velocity Experiment (continued)

Data:
1. 0 seconds; 0 cm
2. 2 seconds; 30 cm
3. 4 seconds; 60 cm
4. 6 seconds; 90 cm
5. 8 seconds; 120 cm
6. 10 seconds; 150 cm

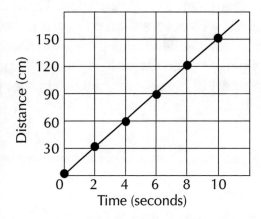

Conclusion: The velocity of the car is 30 cm/2 seconds, or 15 cm/sec.

1. What is the ultimate purpose of a laboratory report, as shown here?

   _____
   _____
   _____
   _____

2. How does the graph add to your understanding of the steps in this process?

   _____
   _____
   _____
   _____

Lesson 20
Assertion and Support

A typical science reading is called an *assertion and support.* In this kind of reading, the author wants to demonstrate a principle, prove a point, or convince you that he or she is correct.

The author has a central point that she or he wishes to prove. This is the assertion, often called the *thesis statement.* In the first paragraph, the author will briefly outline the reasons why the reader should accept the assertion. In the following paragraphs, the author will present each supporting argument. This process is called the support. Often at the end, the author will demonstrate—that is, lead the reader through the steps again to show why the author is correct.

In order to support an assertion, the author will appeal to evidence. This may include experimental data, statistics, observations, expert opinion, facts, and examples. Most authors cite resources, such as journal articles and other research.

Application

Read the following short passage. Then answer the questions that follow.

Did Birds Descend from Dinosaurs?

When you look at a robin in your garden, are you really seeing a miniature *Tyrannosaurus rex*? The evidence seems to say so.

Birds have many characteristics in common with therapods—dinosaurs that walked on two legs. They have claws on their feet for grasping and holding prey. When they walk, birds move their legs in dinosaurlike fashion, bending their legs backwards.

But perhaps the most interesting evidence is the fossil evidence. Several fossils have been found of archaeopteryx, an ancient feathered reptile. Archaeopteryx looks like any small therapod. If only bones had been found, we would believe we had found a little therapod. However, in a few stunning fossils, we can see the obvious outlines of feathers. Unlike modern birds, archaeopteryx also had teeth. Its bones were not hollow, like those of modern birds. *Archaeopteryx* is a transitional species, between dinosaurs and birds.

(continued)

Assertion and Support (continued)

Did Birds Descend from Dinosaurs? (continued)

Based on shared characteristics and fossil evidence, the truth of where birds come from is clear. Our little garden pets are, in fact, related to the most terrifying animals that ever walked the planet.

1. What is the author's assertion?

   _____
   _____
   _____

2. How does the author support this assertion?

   _____
   _____
   _____

3. What resources might an author use to support an assertion?

   _____
   _____
   _____

4. Do you accept the author's assertion? Why or why not?

   _____
   _____
   _____
   _____

Lesson 21
Review

In the following pages, you will find more typical scientific readings. We have discussed many tools to help you improve your understanding of scientific readings. We will apply them to these longer readings. Remember to summarize and paraphrase after every one.

Let's briefly review our reading tools.

Building Vocabulary

Here are tools for building vocabulary.
- Picking up context clues
- Remembering the importance of prefixes, suffixes, and compound words
- Remembering to look for word groups

Before Reading Strategies

Four prereading steps can help you get the most out of reading.
- Previewing
- Predicting
- Using prior knowledge
- Determining the purpose of the passage

Reading Strategies

You learned five reading strategies and graphic organizers.
- **KWL**—What I KNOW, What I WANT to Know, What I LEARNED. Record prior knowledge, results of preview, and what the article taught you.
- **SQ3R**—Survey, Question, Read, Retell, Reflect. Survey an article, ask questions, read, retell it in your own words, and reflect about what you read.
- **Semantic Web.** Draw a "map" of a concept or topic, visually organizing everything you know or read about the concept.
- **Outline.** Write the primary and secondary topics in outline form.
- **Structured Notes.** Determine the organization of a piece, then make a graphic organizer to arrange your notes logically.

Review *(continued)*

Postreading Strategies

- **Summarizing.** You retell the main points of the reading, using the words from the reading.
- **Paraphrasing.** You retell the main points in your own words.

Patterns and Features Common in Science Reading

We looked at five features and patterns often seen in scientific writing.

- **Special terms.** To have a complete understanding, you must understand the scientific vocabulary of the reading.
- **Topic and subtopic.** A common pattern that breaks a large concept into logical parts.
- **Classification.** An organizational technique that defines objects by their membership in one or more groups and subgroups.
- **Steps in a process.** A how-to type of writing that allows the reader to repeat steps.
- **Assertion and support.** The approach science writers use to prove their thesis statements.

PART 6
Practice Readings

READING A

Read the selection, filling in a graphic organizer. When you have finished, summarize or paraphrase the reading on the back of this sheet. Then complete the quiz that follows.

A Year Around the Sun

Many people think that we have winter when we are farthest away from the sun. In fact, however, in the Northern Hemisphere, we are farthest away from the sun in midsummer. So what actually causes the seasons?

We have seasons because the earth is tilted a little on its axis, 23.5 degrees from center. When we have summer, our part of the planet is tilted toward the sun and the sun's rays fall on us most directly. In the winter, we tilt away from the sun. The sun's rays fall at an oblique angle—a more inclined angle. Because of this, the rays of the sun do not warm us as efficiently.

What happens between summer and winter? The Northern Hemisphere is not tilted toward or away from the sun as much, and we have spring and fall. If the earth had no tilt, we would have a springtime world all the time.

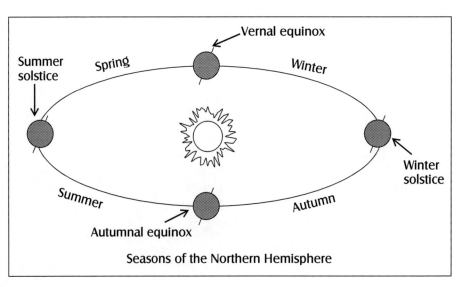

Seasons of the Northern Hemisphere

(continued)

Reading A (continued)

A Year Around the Sun (continued)

You have probably noticed that in the summer we have many long, sunny hours, while in the winter the days are shorter. On the first day of summer, the day is as long as it will ever be, while on the first day of winter the day is as short as it will ever be. These days have special names: *solstice.* The word "solstice" means "sun still."

On one day in the spring and one day in the autumn, day and night are equal. They are each 12 hours long. These days also have a special name: *equinox,* or "equal night." The equinoxes occur on the first day of spring (the vernal equinox) and the first day of autumn (the autumnal equinox).

If we measured the position of the sun in the southern sky at noon on each of those special days, we would find that on the summer solstice, the sun is very high in the southern sky. It also rises and sets much closer to the northern sky than it does at other times of the year. At the winter solstice, the sun is very low in the southern sky. It rises and sets far to the south. At the equinoxes, the sun rises due east and sets due west. It appears to be halfway between the position of the sun at noon on the solstices.

The journey around our sun takes one year. In fact, that is the definition of a year—the amount of time it takes a planet to make one complete trip around the sun. The other planets in our solar system have different years from ours. Those closer to the sun have shorter years. Those farther away from the sun have longer years.

Reading A *(continued)*

QUIZ: A Year Around the Sun

Answer the following. Write your answer in the space provided.

1. Define the following words. Use context clues to help.

    (a) solstice: _____

    _____

    (b) equinox: _____

    _____

    (c) oblique angle: _____

    _____

2. According to the reading, the Northern Hemisphere experiences summer when the earth is farthest from the sun in orbit. Why?

    _____

    _____

    _____

3. If you were looking at the sky, how could you tell what season it is without any other clues?

    _____

    _____

    _____

Read the selection, filling in a graphic organizer. When you have finished, summarize or paraphrase the reading on the back of this sheet. Then complete the quiz that follows.

The Strange Creatures of the Burgess Shale

In British Columbia, Canada, there is a deposit of fossils unlike any on Earth. The creatures of the Burgess Shale lived more than half a billion years ago. Sometime in Earth's early history, an underwater mudslide trapped these animals. This left a perfect fossil record of the great diversity of life at that time. Some of the animals seem to have descendants living today. Others do not seem to be related to any modern animals.

What kind of animals can be seen there? One typical kind of fossil is the trilobite, which had many species. Many were small, but some were the size of large fish! Trilobites were ancient arthropods. They were of the same class as insects, spiders, and crustaceans.

All of the animals in the Burgess Shale were invertebrates—that is, they had no backbone. However, some were chordates. Chordates are animals that have a central nervous system, such as is found in our spinal column. Our ancestors can be seen in the Burgess Shale.

Because the soft mud covered them immediately, many creatures that were not well preserved elsewhere remain intact in the Burgess Shale. Soft body parts that normally decay were imprinted in the mud. This means that scientists can get a good idea of what the animal looked like when it was alive.

The most important thing about the Burgess Shale is that it shows great diversity of life early in Earth's history. By using these fossils, we can try to figure out how life fills the niches available to it in short periods of time.

Reading B (continued)

QUIZ: The Strange Creatures of the Burgess Shale

Answer the following. Write your answer in the space provided.

1. Define the following terms. Use context clues and word parts to help you.

    (a) trilobite: _____
    _____

    (b) invertebrate: _____
    _____

    (c) chordate: _____
    _____

2. Where is the Burgess Shale located today? Where must it have been originally?
    _____
    _____
    _____

3. Why does the Burgess Shale have fossils of some animals that cannot be found anywhere else?
    _____
    _____
    _____

Reading C

Read the selection, filling in a graphic organizer. When you have finished, summarize or paraphrase the reading on the back of this sheet. Then complete the quiz that follows.

What Is a Cell?

All living tissue is made up of cells. Cells were first seen in 1665 by Robert Hooke. He reported that the living plant tissue he looked at under his microscope was made up of many small compartments. Today, using sophisticated electron microscopes, we can see some of the life processes of the living cell.

There are two distinct types of cells. Prokaryotes, which are primitive, lack the diversity of internal structure of the eukaryotes. For example, prokaryotes have no centralized nucleus. The prokaryote's DNA can be found throughout the cell. It contains no organelles. Eukaryotes have highly specialized organelles and other structures. The eukaryote's DNA is enclosed in the nucleus.

Within the prokaryote, cell division can take place from a simple budding-off anywhere on the cell. The following picture, of a bacterium dividing, shows the relative lack of structure in the organism.

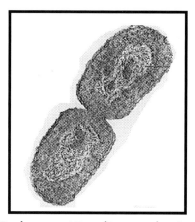

Prokaryote (*E. coli*) reproduction

Eukaryotic cells, however, begin the process of cell division, or mitosis, within the nucleus itself. DNA strands called chromosomes separate. When there are two separate copies of the genetic code, the cell continues to divide. First, the nucleus splits. Finally, the whole cell divides, leaving two daughter cells. The picture on the next page shows mitosis in various stages, including a nonreproducing (resting) cell, the initial stages of reproduction, and final division.

(continued)

Reading C (continued)

What Is a Cell? (continued)

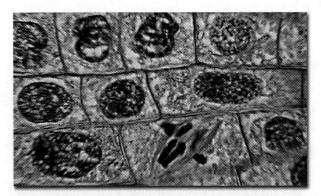

Eukaryotic cell division in onion plant

Note the structure in each cell. Because this is a plant sample, a rigid cell wall separates the cells from one another. This cell wall gives the plant its basic outer structure. In animal cells, the thinner, more pliable cell membrane separates cells from one another. The giant mass in the center of each cell is the nucleus. Within the nucleus, we can see the dividing chromosomes. Outside the nucleus, we can see smaller structures called organelles. These smaller structures are responsible for processing and storing the nutrients, exchanging gas and nutrients, and other important functions. These structures are called the cytoplasm. However, it is the nucleus that runs the show. The nucleus sends chemicals out into the cytoplasm that cause the organelles to do their jobs.

The jobs that the organelles do may vary from day to day. For example, animal cells require a certain amount of dissolved salt to function correctly. The membrane organelles let salt into the cell. However, if the salt level within the cell reaches a proportion dangerous to the cell, the same membrane organelle may become temporarily impermeable to salt. The nucleus regulates this process. This leads some people to liken the nucleus of the cell to the brain of an animal.

Reading C (continued)

QUIZ: What Is a Cell?

Answer the following. Write your answer in the space provided.

1. Define the following terms based on context and word parts.

    (a) prokaryotes: _____

    (b) eukaryotes: _____

    (c) organelles: _____

    (d) cytoplasm: _____

    (e) impermeable: _____

2. Draw a picture of chromosomes dividing within the nucleus of a cell. Label major parts of the cell.

3. What are the main differences between prokaryotes and eukaryotes?

_____

_____

_____

Blank Graphic Organizers

4-P Chart

1. Preview	2. Predict	3. Prior Knowledge	4. Purpose

KWL Chart

K What I KNOW	W What I WANT to Know	L What I LEARNED

SQ3R Chart

S Survey	Q Question	R Read	R Recall	R Reflect

Semantic Web

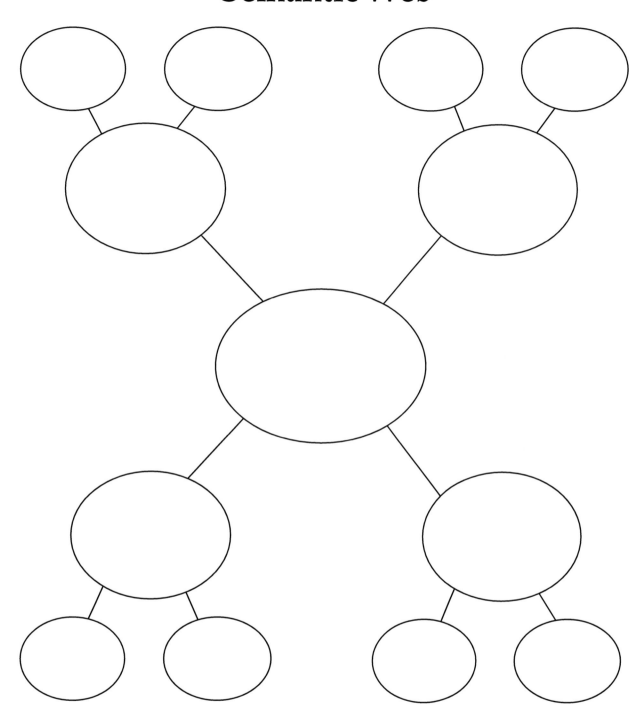

Outline

I. _____

    A. _____

    B. _____

II. _____

    A. _____

    B. _____

III. _____

    A. _____

    B. _____

Structured Notes (some options)

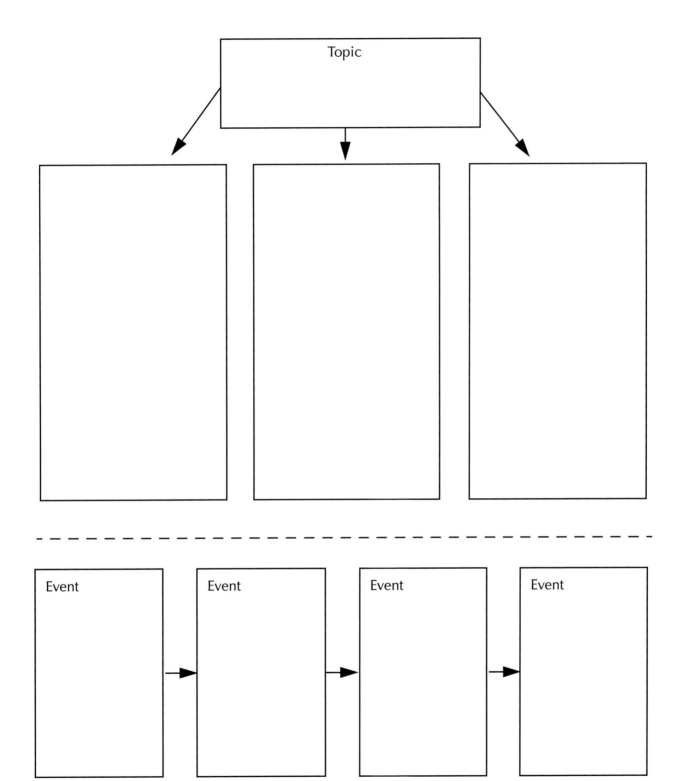

Teacher's Guide and Answer Key

While we provide answers to questions that have specific right or wrong answers, many of the techniques used in this book are subjective and answers will vary.

Part 1: Building Vocabulary

Lesson 1: Using Context Clues

This lesson names some types of context clues students might encounter while reading. The point of the lesson, however, is not to learn the types of clues; rather, students need to recognize a clue when they see one. This is a skill that can be practiced in the classroom during independent reading, or as part of homework assignments, as well as while using this book. Learning to use context clues empowers readers. They can rely largely on themselves, rather than on a reference library or a teacher, to decipher unfamiliar words.

Application

1. *Allosaurus*; definition
2. carnivore; definition
3. herbivore; example/opposite/restatement
4. erect; definition
5. horizontally; opposite

Lesson 2: Word Parts

This lesson focuses on using prefixes, suffixes, and compounds to unlock the meaning of unknown words. Sometimes context may not be enough to understand a difficult word. There are other tools to use, including breaking down words into their components. Some of the prefixes and suffixes are general, while others are specific to scientific reading. Students should not try to memorize these; rather, they should stay alert to the use of suffixes, prefixes, and root words in their reading.

Application

1. body change
2. sending heat away from the body
3. namers of stars
4. a long ridge of sand
5. related to the sun

Lesson 3: Word Groups

This lesson introduces looking for words within other words to help define new words. Some readers may assume they cannot understand a reading with longer words in it. This lesson gives them a way to break down long words into manageable parts.

Application

1. started from; origin, originally
2. sooner or later; event
3. pertaining to Asia
4. those who come after; descend

Part 2: Prereading

Lesson 4: Previewing

Application

Answers will vary, but may include savannah, grassland, watering hole, herd, prey, predator, scavenger.

Lesson 5: Predicting

Application

Answers will vary. Sample answers:
1. The audience is probably not a group of experts. The article is like a nature show, not a news story or a scientific journal.
2. I expect that the story will talk about how things change on the savannah over a year.

Lesson 6: Prior Knowledge

Application

1. Answers will vary, but may include savannah, migration, season, herd, wildebeest, zebra, lion, prey, predator, scavenger, nature.
2. Answers will vary. Sample sentences: Every year on the savannah they have a wet season and a dry season. The animals migrate every year in search of food and water. The prey animals migrate first, followed by the predators. I once went to a zoo where they kept prey herds and predators in the same enclosure, and I saw a pride of lions take down a gazelle.
3. Answers will vary.

Lesson 7: Purpose

Application

1. Answers will vary. Sample answer: The purpose for reading this article is to become familiar with the seasonal changes in a particular ecosystem, the African savannah.
2. Answers will vary. Sample answer: Yes, it is helpful to have a framework in my mind before I start reading. It sets up the reading so I don't have to start the reading cold with no expectations.

Part 3: Reading Strategies

Lesson 8: Introduction to Reading Strategies

1. Answers will vary. Typical readings for the average sixth grader include e-mail, homework assignments, television listings, telephone messages, and so forth.
2. Answers will vary. Typical informational reading passages might include textbook chapters, web sites, research sources, and so forth.

Lesson 9: KWL

Application

Charts will vary. Sample answers:

K—What I KNOW

- A forest is a place where trees grow.
- Trees are plants.
- Other life-forms also live in forests.

W—What I WANT to Know

- What sorts of other plants live in forests?
- What kinds of other living things can be found in a forest?
- Why do animals live in a forest?

L—What I LEARNED

- Fungi are decomposers.
- There are thousands of species of animals in the forest.
- All animals play an important role in the health of a forest.

Quiz

1. (c)
2. wildflowers, shrubs, weeds
3. They break down dead matter.
4. plants
5. Answers will vary.

Lesson 10: SQ3R

Application

Charts will vary. Sample answers:

S—Survey

- Stars are grouped into constellations.
- They are a line-of-sight effect.
- Stars in the path of the sun and moon are zodiac constellations.

Q—Questions

- What are constellations supposed to represent?
- How can stars in a constellation be far apart from one another?
- What is the zodiac?

R—Read

- There are 88 different constellations that represent ancient myths, animals, or modern scientific instruments.
- The universe has depth. Stars in any given constellation can be nearer to the earth or farther away.
- The zodiac is the path the sun and moon and planets seem to take across the sky. The constellations that fall in that path are called the zodiacal constellations.

R—Recall

- There are 2,000 stars you can see with your naked eye. All of these stars are in one constellation or another.
- The stars in a constellation are not necessarily near one another in space. They look like dot-the-dot drawings from Earth.
- When the sun is in a particular zodiac constellation, we say that a person born during that time has that zodiac sign.

R—Reflect

- Why did ancient people group stars this way? What was the purpose?

- Do particular zodiac signs mean anything, in terms of personality?

Quiz
1. (a)
2. (d)
3. in late January and February
4. one year
5. Answers will vary, but should include the idea that the Greeks were the first western astronomers.

Lesson 11: Semantic Web

Application

At the bottom of the page is a sample semantic web, appropriate for sixth graders. Your students' webs may vary greatly.

Quiz
1. (a)
2. (c)
3. predators
4. Dams are hard for the salmon to get over; there is a lot of pollution at river mouths and around harbors, which causes salmon losses.

Lesson 12: Outline

Application

Outlines will vary. Here is a sample student outline.

I. How an eclipse begins
   A. birds quiet
   B. crickets chirp
   C. air cools
   D. sky darkens

II. What is a solar eclipse?
   A. moon lies in direct path between sun and the earth
   B. only happens at new moon phase
   C. only narrow band of the earth will see total eclipse
   D. others will see nothing, or partial eclipse
   E. moon appears to cover face of sun
   F. only corona remains
   G. totality doesn't last long

III. How to watch an eclipse safely
   A. welder's glass
   B. pinhole viewer

Quiz
1. (b)
2. (d)
3. It gets darker, birds stop singing, insects wake up, the air gets cooler.
4. They think it is getting dark because night is falling.
5. because the moon only completely covers the sun from certain angles on the earth

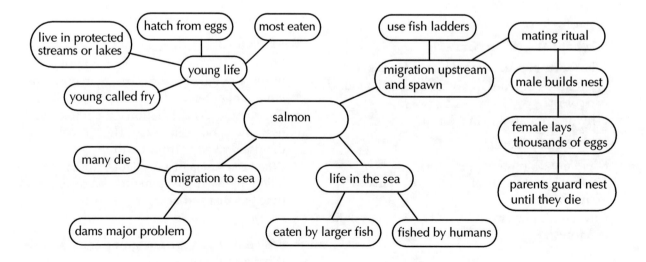

Teacher's Guide and Answer Key **83**

Lesson 13: Structured Notes

Application

Structured notes will vary. Sample answer:

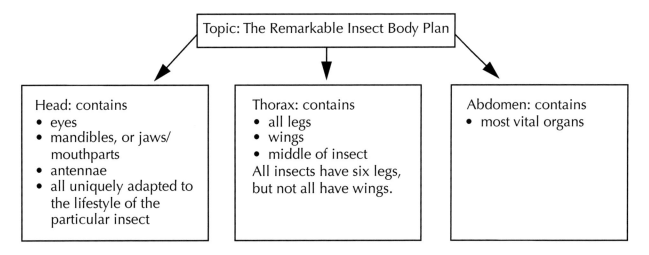

Quiz

1. (c)
2. 3; no exceptions
3. legs and wings
4. antennae, eyes, mandibles
5. Answers will vary, but should include the facts that insects evolve to fill available niches and that the body plan is very versatile.

Part 4: Postreading

Lesson 14: Summary and Paraphrase

Application

Summaries and paraphrasing will vary from student to student. Here are some samples.

Summary:

Volcanic rock is called igneous rock and is formed inside volcanoes. There are several kinds. Basalt is usually gray, but can be pink or black. It will become granite one day. Sometimes basalt forms geodes, where crystals grow. Pumice is light, air-filled rock that can float in water. Obsidian is sometimes called volcanic glass and is hard and shiny. People make arrowheads from it.

Paraphrase:

Another word for volcanic rock is "igneous rock." There are three different kinds in this article. Basalt is one, pumice is another, obsidian is a third. Basalt can form round balls where crystals form, pumice floats in water, and obsidian is used to make arrowheads.

Part 5: Reading in Science

Lesson 15: Common Features and Patterns in Science Reading

List of science readings that students may encounter: science magazine articles; newspaper articles; medication inserts; textbooks; weather forecasts; maps; instructions for using lawn, household, or personal care chemicals

Lesson 16: Special Terms

Application

1. (c); word parts
2. (d); word parts
3. (c); context clues
4. (a); word groups
5. (a); context clues
6. (c); word parts

Lesson 17: Topic and Subtopic

Application

1. life on early earth
2. Bacteria, soft-bodied organisms. They are indicated by subheads and by change in verbal direction.
3. Answers may vary, but should include that the author is giving the reader an idea of his or her opinion by organizing living things in this way. The author explains how life began to diversify, moving from simple to complex.
4. Answers may vary, but should include the idea that life evolves from simple to ever more complex organisms, even up to the extreme complexity of modern organisms.
5. Answers will vary.

Lesson 18: Classification

Application

1. Kingdom: Protoctista
   Phylum: Bacillariophyta
   Class: Centrales
   Order: Thalassiosira
   Family: unknown
   Genus: *Thalassiosira*
   Species: *nordenskjoldii*
2. Answers will vary, but should include radial versus bilateral symmetry.

Lesson 19: Steps in a Process

Application

1. to reproduce the experiment later
2. Answers will vary.

Lesson 20: Assertion and Support

Application

1. Birds descend from dinosaurs.
2. by listing shared characteristics and fossil evidence
3. journal articles, experiments, other evidence
4. Answers will vary.

Part 6: Practice Readings

Reading A: A Year Around the Sun

You may choose a reading strategy for your students or have students select their own. While this reading lends itself to KWL, SQ3R, semantic web, outlining, and structured notes, it is a particularly good reading for KWL.

Quiz

1. (a) solstice: first day of winter or summer; means "sun still"
   (b) equinox: first day of spring or fall; means "equal night"
   (c) oblique angle: means "inclined angle"; results in less of the sun's energy falling directly on the earth
2. because the Northern Hemisphere is tilted toward the sun
3. the height of the sun at noon in the southern sky; where the sun rises and sets

Reading B: The Strange Creatures of the Burgess Shale

You may choose a reading strategy for your students or have students select their own. While this reading lends itself to KWL, SQ3R, semantic web, outlining, and structured notes, it is a particularly good reading for SQ3R or structured notes.

Quiz

1. (a) trilobite: ancient arthropod
   (b) invertebrate: animal without a backbone
   (c) chordate: animal with a central nervous system
2. in British Columbia, Canada; under the sea
3. The soft mud that covered the animals made imprints of them.

Reading C: What Is a Cell?

You may choose a reading strategy for your students or have students select their own. While this reading lends itself to KWL, SQ3R, semantic web, outlining, and structured notes, it is a particularly good reading for semantic webbing or outlining.

Quiz

1. (a) prokaryotes: primitive cells
   (b) eukaryotes: complex cells with a defined nucleus
   (c) organelles: specialized cell parts
   (d) cytoplasm: anything outside the nucleus of a cell
   (e) impermeable: unable to be penetrated
2. Students should be able to label cell wall, cell membrane, nucleus, organelles, chromosomes.
3. Prokaryotes have no nucleus and no organelles; eukaryotes begin the process of cell division and have a defined nucleus and highly specialized organelles.

Share Your Bright Ideas

We want to hear from you!

Your name_____Date_____

School name_____

School address_____

City_____State_____Zip_____Phone number (_____)_____

Grade level(s) taught_____Subject area(s) taught_____

Where did you purchase this publication?_____

In what month do you purchase a majority of your supplements?_____

What moneys were used to purchase this product?

___School supplemental budget   ___Federal/state funding   ___Personal

Please "grade" this Walch publication in the following areas:

	A	B	C	D
Quality of service you received when purchasing	A	B	C	D
Ease of use	A	B	C	D
Quality of content	A	B	C	D
Page layout	A	B	C	D
Organization of material	A	B	C	D
Suitability for grade level	A	B	C	D
Instructional value	A	B	C	D

COMMENTS:_____
_____

What specific supplemental materials would help you meet your current—or future—instructional needs?
_____

Have you used other Walch publications? If so, which ones?_____

May we use your comments in upcoming communications?   ___Yes   ___No

Please **FAX** this completed form to **888-991-5755**, or mail it to

Customer Service, J. Weston Walch, Publisher, P. O. Box 658, Portland, ME 04104-0658

We will send you a **FREE GIFT** in appreciation of your feedback. **THANK YOU!**